Important Astronauts, Cosmonauts, and Other Spaceflight Personalities

Copyright Page

The book is copyrighted for 2020

Important Astronauts, Cosmonauts, and Other Spaceflight
Personalities

The Living in Space Series Book 8

By Martin K. Ettington

All Rights Reserved USA 2020

ISBN: 9798687020336

Printed in the United States of America

Important Astronauts, Cosmonauts, and Other Spaceflight Personalities

This book is the eighth volume of "The Living in Space Series". Each volume focuses on one particular technology of living in space.

This book covers the Astronauts, Cosmonauts, and other individuals who were responsible for some of the biggest space adventures to date.

When I was a kid growing up in the 1960s my friends and I were always listening to the latest news about the Mercury Seven and their latest adventures flying into space.

As of 2020 less than 600 persons have gone into space. These people have pioneered many different types of space activities and are responsible for many legends.

In 1985 while working for Hewlett Packard I got myself transferred to work at the Johnson Space Center in Houston, Texas for a couple of years. It was incredible for me because I got to see the manned space flight center up close, walk inside of Space mockups, and meet many astronauts. In fact it was David Wolf—who later became an astronaut who took me up for rides in his Pitt Special Acrobatic plane (where I got really sick) which convinced me to get my own pilots license. I even applied for the astronaut corps but didn't make it.

I also knew Julie Resnick who was killed in the Challenger explosion. It was very impressive attending the Challenger public funeral where President Reagan spoke.

As a lifelong space buff, I'm very happy to write this book series and introduce the key players here. In this book we will profile those persons who have been the star players in man's exploration of space, and who have made major space contributions over the years.

Important Astronauts, Cosmonauts, and Other Spaceflight Personalities

Other books by Martin K. Ettington

Spiritual and Metaphysics Books:
Prophecy: A History and How to
Guide
God Like Powers and Abilities
Enlightenment for Newbies
Removing Illusions to Find True
    Happiness
Using the Scientific Method to
    Study the Paranormal
A Compendium of Metaphysics
    and How to Guides (Six
    books together in one
    volume)
Love from the Heart
The Enlightenment Experience
Learn Your Soul's Purpose
Pursuing Enlightenment
A Modern Man's Search for Truth
Use Intuition and Prophecy to
    Improve Your Life
The Handbook of Spiritual and
    Energy Healing

Longevity & Immortality:
Physical Immortality: A History
    and How to Guide
The Commentaries of Living
Immortals
Records of Extremely Long Lived
    Persons
Enlightenment and Immortality
Longevity Improvements from
Science
The 10 Principles of Personal
    Longevity
Telomeres & Longevity
The Diets and Lifestyles of the
    Worlds Oldest Peoples
The Longevity Six Books Bundle

Science Fiction:
Out of This Universe
Personal Freedom-Parts 1 & 2
The Psychic Soldier Series:
    Book 1-Himalayan Journey
    Book 2-A Soldier is Born
    Book 3-Fighting For Right
    Book 4-Earth Protector
The Immortality Sci Fi Bundle

The God Like Powers Series:
Human Invisibility
Invulnerability and Shielding
Teleportation
Psychokinesis
Our Energy Body, Auras, and
Thoughtforms

The God Like Powers Series—
    Volume 1 Compilation

The Yoga Discovery Series:
Yoga-An Ancient Art Form
Hatha Yoga-Helping you Live
Better
Raja Yoga-Through the Ages
The Yoga Discovery Package

Business & Coaching Books:
Creating, Paublishing, &
    Marketing Practitioner
    Ebooks
Building a Successful Longevity
    Coaching Business
Why Become a Coach?
The Professional Coaching
Success Trilogy
2020-Make Money Writing and
    Selling Books
The 2020 Handbook of High
    Paying Work Without a
    College Degree

Science, Technology, and Misc.
Future Predictions By and
    Engineer & Seer
The Unusual Science &
    Technology Bundle
The Real Atlantis-In the Eye of
    the Sahara
Are Cryptozoological Animals
    Real or Imaginary?
Real Time Travel Stories From a
    Psychic Engineer
Removing Limits On Our
    Consciousness-And
    Thinking Outside the
    Box

Important Astronauts, Cosmonauts, and Other Spaceflight Personalities

33 Incredible True Survival Stories

How to Survive Anything: From the Wilderness to Man Made Disasters

Building and Stocking a Nuclear Shelter for less than $10,000

All About Mars Journeys and Settlement

Mining the Asteroid Belt

Ancient History

The Real Atlantis-In the Eye of the Sahara

Ancient & Prehistoric Civilizations

Ancient & Prehistoric Civilizations-Book Two

The History of Antediluvian Giants

The Antediluvian History of Earth

Ancient Underground Cities and Tunnels

Strange Objects Which Should Not Exist

Strange and Ancient Places in the USA

A Theory of Ancient Prehistory And Giant Aliens

Aliens and Space

Aliens and Secret Technology

Aliens Are Already Among Us

Designing and Building Space Colonies

Humanity and the Universe

All About Moon Bases

All About Mars Journeys and Settlement

The Space and Aliens Six Books Bundle

A Theory of Ancient Prehistory and Giant Aliens

The Space Colonies and Space Structures Coloring Book

All About Asteroids

Spaceships, Past, Present, and Future

The Longevity Training Series

(A transcription of the online Multimedia Longevity Coaching Training Program)

The Personal Longevity Training Series-Book1-Long Lived Persons
The Personal Longevity Training Series-Book2-Your Soul's Purpose
The Personal Longevity Training Series-Book3-Enable Your Life Urge
The Personal Longevity Training Series-Book4-Your Spiritual Connection
The Personal Longevity Training Series-Book5-Having Love in Your Heart
The Personal Longevity Training Series-Book6-Energy Body Health
The Personal Longevity Training Series-Book7-The Science of Longevity
The Personal Longevity Training Series-Book8-Physical Body Health
The Personal Longevity Training Series-Book9-Avoiding Accidents
The Personal Longevity Training Series-Book10-Implementing These Principles

The Personal Longevity Training Series-Books One Thru Ten

These books are all available in digital and printed formats from my website and on Amazon, Barnes & Noble, Apple ITunes, and many other sites

My Books Website is: http://mkettingtonbooks.com

Signup for our Mailing List to get the following:

1) A discount coupon for 25% discount on all books on our site

2) Occasional Notices of new books available

3) Occasional Email on other offerings of ours (Monthly)

Go to this link to sign-up:

http://personal-longevity.com/mkebooks/emailsignup/

And click this link to get the FREE 102 page Ebook titled "Secrets of Many Things"

If you have any questions about this book or other subjects please contact the Author at:

mke@mkettingtonbooks.com

Important Astronauts, Cosmonauts, and Other Spaceflight Personalities

Table of Contents

1.0 Introduction

This book is the eighth in "The Living in Space Series". Each book covers a specific focus of Space Travel and/or technology.

This book covers the Astronauts, Cosmonauts, and other individuals who were responsible for some of the biggest space adventures to date.

When I was a kid growing up in the 1960s my friends and I were always listening to the latest news about the Mercury Seven and their latest adventures flying into space.

As of 2020 less than 600 persons have gone into space. These people have pioneered many different types of space activities and are responsible for many legends.

In 1985 while working for Hewlett Packard I got myself transferred to work at the Johnson Space Center in Houston, Texas for a couple of years. It was incredible for me because I got to see the manned space flight center up close, walk inside of Space Shuttle mockups, and meet many astronauts. In fact it was David Wolf—who later became an astronaut who took me up for rides in his Pitt Special Acrobatic plane (where I got really sick) which convinced me to get my own pilots license. I even applied for the astronaut corps but didn't make it.

I also knew Julie Resnick who was killed in the Challenger explosion. It was very impressive attending the Challenger public funeral where President Reagan spoke.

As a lifelong space buff, I'm very happy to write this book series and introduce the key players here. In this book we will profile those persons who have been the star players in man's exploration of space, and who have made major space contributions over the years.

2.0 Space Leaders and Visionaries

I've included visionary and famous engineers in this chapter because they were architects of and pioneers in our manned in space ventures. President Kennedy is also included because of his famous speech about man landing on the moon which set the tone for a decade of manned space efforts.

2.1 Werner Von Braun

Wernher Magnus Maximilian Freiherr von Braun (March 23, 1912 – June 16, 1977) was a German–born American aerospace engineer and space architect. He was the leading figure in the development of rocket technology in Nazi Germany and a pioneer of rocket and space technology in the United States.

While in his twenties and early thirties, von Braun worked in Nazi Germany's rocket development program. He helped design and develop the V-2 rocket at Peenemünde during World War II. Following the war he was secretly moved to the United States, along with about 1,600 other

German scientists, engineers, and technicians, as part of Operation Paperclip. He worked for the United States Army on an intermediate-range ballistic missile program, and he developed the rockets that launched the United States' first space satellite Explorer 1.

In 1960, his group was assimilated into NASA, where he served as director of the newly formed Marshall Space Flight Center and as the chief architect of the Saturn V super heavy-lift launch vehicle that propelled the Apollo spacecraft to the Moon. In 1967, von Braun was inducted into the National Academy of Engineering, and in 1975, he received the National Medal of Science. He advocated for a human mission to Mars.

2.2 Sergei Korolev

Sergei Pavlovich Korolev (also transliterated as Sergey Pavlovich Korolyov; January 12 1907 – 14 January 1966) was a lead Soviet rocket engineer and spacecraft designer during the Space Race between the United States and the Soviet Union in the 1950s and 1960s. He is regarded by many as the father of practical astronautics. He was involved in the development of the R-7 Rocket, Sputnik 1, and launching Laika and the first human being, Yuri Gagarin, into space.

Although Korolev trained as an aircraft designer, his greatest strengths proved to be in design integration, organization and strategic planning. Arrested on a false official charge as a "member of an anti-Soviet counter-revolutionary organization" (which would later be reduced to "saboteur of military technology"), he was imprisoned in 1938 for almost six years, including some months in a Kolyma labor camp. Following his release he became a recognized rocket designer and a key figure in the development of the Soviet Intercontinental ballistic missile program. He later directed the Soviet space program and was made a Member of Soviet Academy of Sciences, overseeing the early successes of the Sputnik and Vostok projects including the first human Earth orbit mission by Yuri Alexeyvich Gagarin on 12 April 1961. Korolev's unexpected death in 1966 interrupted implementation of his plans for a Soviet crewed Moon landing before the United States 1969 mission.

Before his death he was officially identified only as Glavny Konstruktor, or the Chief Designer, to protect him from possible cold war assassination attempts by the United States. Even some of the cosmonauts who worked with him were unaware of his last name; he only went by Chief Designer. Only following his death in 1966 was his identity revealed and he received the appropriate public recognition as the driving force behind Soviet accomplishments in space exploration during and following the International Geophysical Year.

2.3 President Kennedy's Moon Speech

"We choose to go to the Moon", officially titled as the Address at Rice University on the Nation's Space Effort, is a speech delivered by United States President John F. Kennedy about the effort to reach the Moon to a large crowd gathered at Rice Stadium in Houston, Texas, on September 12, 1962. The speech, largely written by Kennedy advisor and speechwriter Ted Sorensen, was intended to persuade the American people to support the Apollo program, the national effort to land a man on the Moon.

In his speech, Kennedy characterized space as a new frontier, invoking the pioneer spirit that dominated American folklore. He infused the speech with a sense of urgency and destiny, and emphasized the freedom enjoyed by Americans to choose their destiny rather than have it chosen for them. Although he called for competition with the Soviet Union, Kennedy also proposed making the Moon landing a joint project.

The speech resonated widely and is still remembered, although at the time there was disquiet about the cost and value of the Moon-landing effort. Kennedy's goal was realized, in July 1969, with the successful Apollo 11 mission.

2.4 Elon Musk

Elon Reeve Musk FRS (born June 28, 1971) is a business magnate, industrial designer, engineer, and philanthropist. He is the founder, CEO, CTO and chief designer of SpaceX; early investor, CEO and product architect of Tesla, Inc.; founder of The Boring Company; co-founder of Neuralink; and co-founder and initial co-chairman of OpenAI. He was elected a Fellow of the Royal Society (FRS) in 2018. In 2018, he was ranked 25th on the Forbes list of The World's Most Powerful People, and was ranked joint-first on the Forbes list of the Most Innovative Leaders of 2019. As of September 2, 2020, his net worth was estimated by Forbes to be US$93.3 billion, making him the 5th richest person in the world. He is the longest tenured CEO of any automotive manufacturer globally.

Musk was born to a Canadian mother and South African father and raised in Pretoria, South Africa. He briefly attended the University of Pretoria before moving to Canada when he was 17 to attend Queen's University. He transferred to the University of Pennsylvania two years later, where he received dual bachelor's degrees in economics and physics. He moved to California in 1995 to begin a Ph.D. in applied physics and material sciences at

Stanford University, but decided to pursue a business career instead of enrolling. With his brother Kimbal, he co-founded Zip2, a web software company, which was acquired by Compaq for $307 million in 1999. Musk then founded X.com, an online bank. It merged with Confinity in 2000, which had launched PayPal the previous year and was subsequently bought by eBay for $1.5 billion in October 2002.

In May 2002, Musk founded SpaceX, an aerospace manufacturer and space transport services company, of which he is CEO and lead designer. He joined Tesla Motors, Inc. (now Tesla, Inc.), an electric vehicle manufacturer, in 2004, the year after it was founded, became its product architect, and became its CEO in 2008. In 2006, he helped create SolarCity, a solar energy services company (now a subsidiary of Tesla). In 2015, Musk co-founded OpenAI, a nonprofit research company that aims to promote friendly artificial intelligence. In July 2016, he co-founded Neuralink, a neurotechnology company focused on developing brain–computer interfaces. In December 2016, Musk founded The Boring Company, an infrastructure and tunnel construction company focused on tunnels optimized for electric vehicles. In addition to his primary business pursuits, Musk envisioned an open-source high-speed transportation system known as the Hyperloop based on the concept of a vactrain.

3.0 First Men in Space

The first men in space—both Americans and Russians have a big place in history. These people risked their lives on man's first efforts to reach outside of Earth's atmosphere.

3.1 Yuri Gagarin-Soviet Union

Yuri Alekseyevich Gagarin (9 March 1934 – 27 March 1968) was a Soviet Air Forces pilot and cosmonaut who became the first human to journey into outer space, achieving a major milestone in the Space Race; his capsule, Vostok 1, completed one orbit of Earth on 12 April 1961. Gagarin became an international celebrity and was awarded many medals and titles, including Hero of the Soviet Union, his nation's highest honor.

Born in the village of Klushino near Gzhatsk (a town later renamed after him), in his youth Gagarin was a foundryman at a steel plant in Lyubertsy. He later joined the Soviet Air Forces as a pilot and was stationed at the Luostari Air Base, near the Norwegian border, before his selection for the Soviet space program with five other cosmonauts. Following his spaceflight, Gagarin became deputy training director of the Cosmonaut Training Centre, which was later named after him. He was also elected as a deputy of the Soviet of the Union in 1962 and then to the Soviet of Nationalities, respectively the lower and upper chambers of the Supreme Soviet.

Vostok 1 was Gagarin's only spaceflight but he served as the backup crew to the Soyuz 1 mission, which ended in a fatal crash, killing his friend and fellow cosmonaut Vladimir Komarov. Fearing for his life, Soviet officials permanently banned Gagarin from further spaceflights. After completing training at the Zhukovsky Air Force Engineering Academy on 17 February 1968, he was allowed to fly regular aircraft. Gagarin died five weeks later when the MiG-15 training jet he was piloting with his flight instructor Vladimir Seryogin crashed near the town of Kirzhach.

3.2 Alan Shepard-USA

Rear Admiral Alan Bartlett Shepard Jr. (November 18, 1923 – July 21, 1998) was an American astronaut, naval aviator, test pilot, and businessman. In 1961, he became the first American to travel into space, and in 1971, he walked on the Moon.

A graduate of the United States Naval Academy at Annapolis, Shepard saw action with the surface navy during World War II. He became a naval aviator in 1946, and a test pilot in 1950. He was selected as one of the original NASA Mercury Seven astronauts in 1959, and in May 1961 he made the first crewed Project Mercury flight, Mercury-Redstone 3, in a spacecraft he named Freedom 7. His craft entered space, but was not capable of achieving orbit. He became the second person, and the first American, to travel into space, and the first space traveler to manually control the orientation of his craft. In the final stages of Project Mercury, Shepard was scheduled to pilot the Mercury-Atlas 10 (MA-10), which was planned as a three-day mission. He named Mercury Spacecraft 15B Freedom 7 II in honor of his first spacecraft, but the mission was canceled.

Shepard was designated as the commander of the first crewed Project Gemini mission, but was grounded in 1963 due to Ménière's disease, an inner-ear ailment that caused episodes of extreme dizziness and nausea. This was surgically corrected in 1969, and in 1971, Shepard commanded the Apollo 14 mission, piloting the Apollo Lunar Module Antares to the most accurate landing of the Apollo missions. At age 47, he became the fifth, the oldest, and the earliest-born person to walk on the Moon, and the only one of the Mercury Seven astronauts to do so. During the mission, he hit two golf balls on the lunar surface.

Shepard was Chief of the Astronaut Office from November 1963 to July 1969 (the approximate period of his grounding), and from June 1971 until his retirement from the United States Navy and NASA on August 1, 1974. He was promoted to rear admiral on August 25, 1971, the first astronaut to reach that rank.

3.3 John Glenn-USA

John Herschel Glenn Jr. (July 18, 1921 – December 8, 2016) was a United States Marine Corps aviator, engineer, astronaut, businessman and politician. He was the first American to orbit the Earth, circling it three times in 1962. Following his retirement from NASA, he served from 1974 to 1999 as a Democratic United States Senator from Ohio; in 1998, he flew into space again at age 77.

Before joining NASA, Glenn was a distinguished fighter pilot in World War II, China and Korea. He shot down three MiG-15s, and was awarded six Distinguished Flying Crosses and eighteen Air Medals. In 1957, he made the first supersonic transcontinental flight across the United States. His on-board camera took the first continuous, panoramic photograph of the United States.

He was one of the Mercury Seven, military test pilots selected in 1959 by NASA as the nation's first astronauts. On February 20, 1962, Glenn flew the Friendship 7 mission, becoming the first American to orbit the Earth, and the fifth person and third American in space. He received the NASA Distinguished Service Medal in 1962, the Congressional Space Medal of Honor in 1978, was inducted into the U.S. Astronaut Hall of Fame in 1990, and received the Presidential Medal of Freedom in 2012.

Glenn resigned from NASA in January 1964. A member of the Democratic Party, Glenn was first elected to the Senate in 1974 and served for 24 years, until January 1999. In 1998, while still a sitting senator, Glenn flew on Space Shuttle Discovery's STS-95 mission, making him, at age 77, the oldest person to fly in space and the only person to fly in both the Mercury and the Space Shuttle programs. Glenn, both the oldest and the last surviving member of the Mercury Seven, died at the age of 95 in 2016.

3.4 X-15 Pilots-USA

The first non-NASA pilot to be awarded astronaut wings was X-15 pilot R. White, as described in Life Magazine (Aug 3 1962):

Major Bob White of the US Air Force is the nation's newest space hero. He has a brand-new award on his chest that makes him a member of the nation's most exclusive club. It was a special set of pilot's wings that signified he had flown higher than 50 miles above the earth and thereby had qualified as a spaceman.

White flew to 95 km on 1962 Jul 17. If a limit of 100 km instead of 80 km is used, White, Robert Rushworth, Jack McKay, Bill Dana and Mike Adams lose their space traveler status (Joe Engle keeps his because he later flew on Shuttle, and Joe Walker passed the Karman line on his X-15 flights).

Flight 90 of the North American X-15 was a test flight conducted by NASA and the US Air Force in 1963. It was the first of two X-15 missions that passed the 100-km high Kármán line, the FAI definition of space, along with Flight

91 the next month. The X-15 was flown by Joseph A. Walker, who flew both X-15 spaceflights.

3.5 Ghermon Titov-Soviet Union

Gherman Stepanovich Titov 11 September 1935 – 20 September 2000) was a Soviet cosmonaut who, on 6 August 1961, became the second human to orbit the Earth, aboard Vostok 2, preceded by Yuri Gagarin on Vostok 1. He was the fourth person in space, counting suborbital voyages of US astronauts Alan Shepard and Gus Grissom.

Titov's flight finally proved that humans could live and work in space. He was the first person to orbit the Earth multiple times (a total of 17), the first to pilot a spaceship and to spend more than a day in space. He was also the first to sleep in orbit and to suffer from space sickness (becoming the first person to vomit in space).

Titov made the first manual photographs from orbit, thus setting a record for modern space photography. He also was the first person to film the Earth using a professional quality Konvas-Avtomat movie camera, which he used for ten minutes. A month short of 26 years old at launch, he remains the youngest person to fly in space.

In his subsequent life Titov continued to work for the Soviet space program, and played a major role in the Spiral project where he trained to become the first pilot of an orbital spaceplane. However, after the death of Yuri Gagarin in a military aircraft accident in 1968, the Soviet government decided it could not afford to lose its second cosmonaut, and so Titov's career as test pilot ended.

Titov served in the Soviet Air Force, attaining the rank of colonel-general. In his final years in post-Soviet Russia he became a Communist politician. Despite having been chosen second, after Gagarin, to fly into space, it was Titov who later proposed the Soviet Government regularly celebrate Cosmonautics Day on April 12, the day of Gagarin's flight.

3.6 Valentina Tereshkova-Soviet Union

Valentina Vladimirovna Tereshkova (born 6 March 1937) is a member of the Russian State Duma, engineer, and former cosmonaut. She is the first and youngest woman to have flown in space with a solo mission on the Vostok 6 on 16 June 1963. She orbited the Earth 48 times, spent almost three days in space, and remains the only woman to have been on a solo space mission.

Before her selection for the Soviet space program, Tereshkova was a textile factory worker and an amateur skydiver. She joined the Air Force as part of the Cosmonaut Corps and was commissioned as an officer after completing her training. After the dissolution of the first group of female cosmonauts in 1969, Tereshkova remained in the space program as a cosmonaut instructor. She later graduated from the Zhukovsky Air Force Engineering Academy and re-qualified for spaceflight but never went to space again. She retired from the Air Force in 1997 having attained the rank of major general.

Tereshkova was a prominent member of the Communist Party of the Soviet Union, holding various political offices including being a member of the Presidium of the Supreme Soviet from 1974 to 1989. She remained politically active following the collapse of the Soviet Union but twice lost elections to the national State Duma in 1995 and 2003. Tereshkova was later elected in 2008 to her regional parliament, the Yaroslavl Oblast Duma. In 2011, she was elected to the national State Duma as a member of the United Russia party and re-elected in 2016.

4.0 Mercury, Gemini, and Apollo Programs

The three main space programs of the USA in the 1960s were all oriented to providing the experience and technological base to land men on the moon. The Soviets had their own space program and even built a moon rocket but were never successful in landing men on the moon.

4.1 The Mercury Seven

The Mercury Seven were the group of seven astronauts selected to fly spacecraft for Project Mercury. They are also referred to as the Original Seven and Astronaut Group 1. Their names were publicly announced by NASA on April 9, 1959. These seven original American astronauts were Scott Carpenter, Gordon Cooper, John Glenn, Gus Grissom, Wally Schirra, Alan Shepard, and Deke Slayton. The Mercury Seven created a new profession in the United States, and established the image of the American astronaut for decades to come.

All of the Mercury Seven eventually flew in space. They piloted the six spaceflights of the Mercury program that had an astronaut on board from May 1961 to May 1963, and members of the group flew on all of the NASA human spaceflight programs of the 20th century – Mercury, Gemini, Apollo, and the Space Shuttle. Shepard became the first American to enter space in 1961, and later walked on the Moon on Apollo 14 in 1971. Grissom flew Mercury and Gemini missions, but died in 1967 in the Apollo 1 fire; the others all survived past retirement from service. Schirra flew Apollo 7, the first crewed Apollo mission, in Grissom's place. Slayton, grounded with an atrial fibrillation, ultimately flew on the Apollo–Soyuz Test Project in 1975. Glenn became the first American in orbit in 1962, and flew on the Space Shuttle Discovery in 1998 to become, at age 77, the oldest person to fly in space. He was the last living member of the Mercury Seven when he died in 2016 at the age of 95.

4.2 Gemini Astronauts

The Gemini astronauts were pilots who flew in Project Gemini, NASA's second human spaceflight program, between projects Mercury and Apollo. Carrying two astronauts at a time, a senior Command Pilot and a junior Pilot, the Gemini spacecraft was used for ten crewed missions. Sixteen astronauts flew on these missions, with four flying twice.

Gemini was the second phase in the United States space program's larger goal of "landing a man on the Moon and returning him safely to the Earth" before the end of the 1960s, as proposed by President John F. Kennedy. As an intermediary step, Gemini afforded its astronauts the opportunity to gain critical spaceflight experience, performing tasks required in the later Apollo program which fulfilled this objective. Such tasks included rendezvous or station-keeping with other craft, docking, habitation in space over the course of several days, and flying spacecraft with more than one crew member. Importantly, most individuals who flew as Gemini astronauts returned to space as key personnel in the Apollo program, bringing with them their first-hand experience of the operations carried out during Gemini. Among the Gemini astronauts, six later walked on the Moon, another five flew to the Moon without landing, and two participated in Low Earth orbit Apollo missions. Gus Grissom and Ed White were killed in the Apollo 1 disaster, and former Mercury astronaut Gordon Cooper did not perform any further spaceflights.

All Gemini astronauts–excluding the Mercury Seven astronauts already included–were inducted into the U.S. Astronaut Hall of Fame in 1993.

Astronaut participation in Project Gemini was also a strong predictor for future achievement during the Apollo Program:

Every Apollo mission commander, including Gus Grissom and with the exception of Alan Shepard, was a Gemini veteran.

All three crew members of Apollo 11, the first lunar landing-Neil Armstrong, Michael Collins and Buzz Aldrin-were Gemini veterans.

All three of the men who flew to the Moon twice-Jim Lovell, John Young and Gene Cernan-were Gemini veterans. With the exception of Elliot See, every member of NASA's second Astronaut Group—the class of nine men selected following the Mercury Seven—flew as a Gemini astronaut.

4.3 Apollo Astronauts

The Apollo Astronauts tested and rode the Saturn V and Apollo Command modules to land on the Moon. Thus completing one of the biggest quests in history.

4.3.1 Neil Armstrong

Neil Alden Armstrong (August 5, 1930 – August 25, 2012) was an American astronaut and aeronautical engineer, and the first person to walk on the Moon. He was also a naval aviator, test pilot, and university professor.

A graduate of Purdue University, Armstrong studied aeronautical engineering; his college tuition was paid for by the U.S. Navy under the Holloway Plan. He became a midshipman in 1949 and a naval aviator the following year. He saw action in the Korean War, flying the Grumman F9F Panther from the aircraft carrier USS Essex. In September 1951, while making a low bombing run, Armstrong's aircraft was damaged when it collided with an anti-aircraft cable, strung across a valley, which cut off a large portion of one wing. Armstrong was forced to bail out. After the war, he completed his bachelor's degree at Purdue and became a test pilot at the National Advisory Committee for Aeronautics (NACA) High-Speed Flight Station at Edwards Air Force Base in California. He was the project pilot on Century Series fighters and flew the North American X-15 seven times. He was also a participant in the U.S. Air Force's Man in Space Soonest and X-20 Dyna-Soar human spaceflight programs.

Armstrong joined the NASA Astronaut Corps in the second group, which was selected in 1962. He made his first spaceflight as command pilot of Gemini 8 in March 1966, becoming NASA's first civilian astronaut to fly in space. During this mission with pilot David Scott, he performed the first docking of two spacecraft; the mission was aborted after Armstrong used some of his re-entry control fuel to stabilize a dangerous roll caused by a stuck thruster. During training for Armstrong's second and last spaceflight as commander of Apollo 11, he had to eject from the Lunar Landing Research Vehicle moments before a crash.

On July 20, 1969, Armstrong and Apollo 11 Lunar Module (LM) pilot Buzz Aldrin became the first people to land on the Moon, and the next day they spent two and a half hours outside the Lunar Module Eagle spacecraft while Michael Collins remained in lunar orbit in the Apollo Command Module Columbia. When Armstrong stepped onto the lunar surface, he famously said: "That's one small step for [a] man, one giant leap for mankind." Along with Collins and Aldrin, Armstrong was awarded the Presidential Medal of Freedom by President Richard Nixon. President Jimmy Carter presented Armstrong with the Congressional Space Medal of Honor in 1978, and Armstrong and his former crewmates received a Congressional Gold Medal in 2009.

After he resigned from NASA in 1971, Armstrong taught in the Department of Aerospace Engineering at the University of Cincinnati until 1979. He served on the Apollo 13 accident investigation and on the Rogers Commission, which investigated the Space Shuttle Challenger disaster. He acted as a spokesman for several businesses and appeared in advertising for the automotive brand Chrysler starting in January 1979.

Important Astronauts, Cosmonauts, and Other Spaceflight Personalities

4.3.2 Buzz Aldrin

Buzz Aldrin born Edwin Eugene Aldrin Jr., January 20, 1930 is an American engineer, and former astronaut and fighter pilot. Aldrin made three spacewalks as pilot of the 1966 Gemini 12 mission, and as the lunar module pilot on the 1969 Apollo 11 mission, he and mission commander Neil Armstrong were the first two humans to land on the Moon.

Born in Glen Ridge, New Jersey, Aldrin graduated third in the class of 1951 from the United States Military Academy at West Point, with a degree in mechanical engineering. He was commissioned into the United States Air Force, and served as a jet fighter pilot during the Korean War. He flew 66 combat missions and shot down two MiG-15 aircraft.

After earning a Sc.D. degree in astronautics from the Massachusetts Institute of Technology, Aldrin was selected as a member of NASA's Astronaut Group 3, making him the first astronaut with a doctoral degree. His doctoral thesis was Line-of-Sight Guidance Techniques for Manned Orbital Rendezvous, earning him the nickname "Dr. Rendezvous" from fellow astronauts. His first space flight was in 1966 on Gemini 12 during which he spent over five hours on extravehicular activity. Three years later, Aldrin set foot on the Moon at 03:15:16 on July 21, 1969 (UTC), nineteen minutes after Armstrong first touched the surface, while command module pilot Michael Collins remained in lunar orbit. A Presbyterian elder, Aldrin became the first person to hold a religious ceremony on the Moon when he privately took communion.

Upon leaving NASA in 1971, Aldrin became Commandant of the U.S. Air Force Test Pilot School. He retired from the Air Force in 1972, after 21 years of service. His autobiographies Return to Earth (1973), and Magnificent Desolation (2009), recount his struggles with clinical depression and alcoholism in the years after leaving NASA. He continued to advocate for space exploration, particularly a human mission to Mars, and developed the Aldrin cycler, a special spacecraft trajectory that makes travel to Mars more efficient in regard to time and propellant. He has been accorded numerous honors, including the Presidential Medal of Freedom in 1969, and is listed in several Halls of Fame.

4.2.3 Jim Lovell

James Arthur Lovell Jr. born March 25, 1928 is an American retired astronaut, naval aviator, and mechanical engineer. In 1968, as command module pilot of Apollo 8, he became one of the first three humans to fly to and orbit the Moon. He then commanded the 1970 Apollo 13 lunar mission which, after a critical failure en route, circled around the Moon and returned safely to Earth through the efforts of the crew and mission control.

Lovell had previously flown on two Gemini missions, Gemini 7 in 1965 and Gemini 12 in 1966. He was the first person to fly into space four times.

One of 24 people to have flown to the Moon, Lovell was the first person to fly to it twice. He is a recipient of the Congressional Space Medal of Honor and the Presidential Medal of Freedom (in 1970, as one of 17 recipients in the Space Exploration group), and co-author of the 1994 book Lost Moon, on which the 1995 film Apollo 13 was based.

4.2.4 Pete Conrad

I met Pete Conrad at a party once while working at Nasa in Houston and mentioned some of his colleagues from the Skylab mission. He had some funny stories about them.

Charles "Pete" Conrad Jr. (June 2, 1930 – July 8, 1999) was an American NASA astronaut, aeronautical engineer, naval officer and aviator, test pilot, and commanded the Apollo 12 space mission, on which he became the third man to walk on the Moon. Conrad was selected in NASA's second astronaut class in 1962.

Before becoming an astronaut, Conrad earned his bachelor's degree in Aeronautical Engineering from Princeton University—being the first Ivy League astronaut—and joined the U.S. Navy. In 1954 he received his naval aviator wings, served as a fighter pilot and, after graduating from the U.S. Naval Test Pilot School (Class 20), as a project test pilot. In 1959 he was an astronaut candidate for Project Mercury.

He set an eight-day space endurance record in 1965 along with his Command Pilot Gordon Cooper on his first spaceflight, Gemini 5. Later, Conrad commanded Gemini 11 in 1966, and Apollo 12 in 1969. After Apollo, he commanded Skylab 2, the first crewed Skylab mission, in

1973. On the mission, he and his crewmates repaired significant launch damage to the Skylab space station. For this, President Jimmy Carter awarded him the Congressional Space Medal of Honor in 1978.

After he retired from NASA and the Navy in 1973, he became a vice president of American Television and Communications Company. He went on to work for McDonnell Douglas, as a vice president. During his tenure, he served as vice president of marketing, senior vice president of marketing, staff vice president of international business development, and vice president of project development.

Conrad died on July 8, 1999, from internal injuries sustained in a motorcycle accident.

5.0 Space Shuttle and Soyuz/MIR

The era of the Space Shuttle and the Soviet MIR space station brought new crops of Astronauts and Cosmonauts to the forefront of space exploration. Here are some of the better known people from that time.

5.1 John Young

John Watts Young (September 24, 1930 – January 5, 2018) was an American astronaut, naval officer and aviator, test pilot, and aeronautical engineer. He became the ninth person to walk on the Moon as Commander of the Apollo 16 mission in 1972. Young enjoyed the longest career of any astronaut, becoming the first person to fly six space missions over the course of 42 years of active NASA service. He is the only person to have piloted and commanded four different classes of spacecraft: Gemini, the Apollo Command and Service Module, the Apollo Lunar Module, and the Space Shuttle.

Before becoming an astronaut, Young received his Bachelor of Science degree with highest honors in Aeronautical Engineering from the Georgia Institute of

Technology and joined the U.S. Navy. After serving at sea during the Korean War he became a naval aviator, and graduated from the U.S. Naval Test Pilot School (Class 23), setting several world time-to-climb records as a test pilot. Young left the Navy in 1976 with the rank of captain.

In 1965 Young flew on the first crewed Gemini mission, and then commanded the 1966 Gemini 10 mission. In 1969 during Apollo 10, he became the first person to fly solo around the Moon. He then walked on the Moon and drove the Lunar Roving Vehicle on the Moon's surface during Apollo 16, and is one of only three people to have flown to the Moon twice.

Young also commanded two flights of Space Shuttle Columbia: STS-1 in 1981, the Space Shuttle program's first launch, and STS-9 in 1983. Young served as Chief of the Astronaut Office from 1974 to 1987, and retired from NASA in 2004.

5.2 Sally Ride

Sally Kristen Ride (May 26, 1951 – July 23, 2012) was an American astronaut and physicist. Born in Los Angeles, she joined NASA in 1978 and became the first American woman in space in 1983. Ride was the third woman in space overall, after USSR cosmonauts Valentina Tereshkova (1963) and Svetlana Savitskaya (1982). Ride remains the youngest American astronaut to have traveled to space, having done so at the age of 32. After flying twice on the Orbiter Challenger, she left NASA in 1987.

Ride worked for two years at Stanford University's Center for International Security and Arms Control, then at the University of California, San Diego as a professor of physics, primarily researching nonlinear optics and Thomson scattering. She served on the committees that investigated the Challenger and Columbia Space Shuttle disasters, the only person to participate in both. Ride died of pancreatic cancer on July 23, 2012.

Important Astronauts, Cosmonauts, and Other Spaceflight Personalities

5.3 Bruce McCandless II

Bruce McCandless was the first man to free fly in his spacesuit outside of a spacecraft using the Manned Maneuvering Unit. (MMU)

At the age of 28, McCandless was selected as the youngest member of NASA Astronaut Group 5 (labelled the "Original Nineteen" by John W. Young) in April 1966. According to space historian Matthew Hersch, McCandless and Group 5 colleague Don L. Lind were "effectively treated ... as scientist-astronauts" (akin to those selected in the fourth and sixth groups) by NASA due to their substantial scientific experience, an implicit reflection of their lack of the test pilot experience highly valued by Deke Slayton and other NASA managers at the time; this would ultimately delay their progression in the flight rotation.

He served as mission control capsule communicator (CAPCOM) on Apollo 11 during the launch and during the first lunar moonwalk (EVA) by Neil Armstrong and Buzz

Aldrin before joining the astronaut support crew for the Apollo 14 mission, on which he doubled as a CAPCOM. Thereafter, McCandless was reassigned to the Skylab program, where he received his first crew assignment as backup pilot for the space station's first crewed mission alongside backup commander Rusty Schweickart and backup science pilot Story Musgrave. Following this assignment, he again served as a CAPCOM on Skylab 3 and Skylab 4. Notably, McCandless was a co-investigator on the M-509 astronaut maneuvering unit experiment that was flown on Skylab; this eventually led to his collaboration on the development of the Manned Maneuvering Unit (MMU) used during Space Shuttle EVAs. Although he was classified as a Shuttle pilot until 1983, McCandless ultimately chose to work on the MMU as a mission specialist due to the prestige of the program (which ensured a flight assignment) and his lack of test pilot experience.

He was responsible for crew inputs to the development of hardware and procedures for the Inertial Upper Stage (IUS), Hubble Space Telescope, the Solar Maximum Repair Mission, and the International Space Station program.

McCandless logged over 312 hours in space, including four hours of MMU flight time. He flew as a mission specialist on STS-41-B and STS-31.

5.4 Story Musgrave

Story Musgrave was one of the crew who worked on repairing the Hubble Space Telescope. This was a major save and success for NASA.

Education

Musgrave attended Dexter School in Brookline, Massachusetts and St. Mark's School in Southborough, Massachusetts from 1947 to 1953. He dropped out of St. Mark's in his senior year when a car accident "caused him to miss a substantial amount of vital pre-graduation exam schooling."

While serving in the Marines, he completed his GED. Following his discharge, Musgrave received a B.S. in mathematics and statistics from Syracuse University in 1958. He went on to receive an M.B.A. in operations analysis and computer programming from the University of California, Los Angeles in 1959, a B.A. in chemistry from Marietta College in 1960, an M.D. degree from Columbia University College of Physicians and Surgeons in 1964, an

M.S. in physiology and biophysics from the University of Kentucky in 1966, and a M.A. in literature from the University of Houston–Clear Lake in 1987.

NASA career

Musgrave was selected as a scientist-astronaut by NASA in August 1967 as a member of NASA Astronaut Group 6. After completing flight and academic training, he worked on the design and development of the Skylab Program. In 1973, he was the backup Science Pilot for Skylab 2, becoming the first Group 6 astronaut to receive a potential flight assignment.

Musgrave participated in the design and development of all Space Shuttle extra-vehicular activity equipment, including spacesuits, life support systems, airlocks and Manned Maneuvering Units. From 1979 to 1982, and 1983 to 1984, he was assigned as a test and verification pilot in the Shuttle Avionics Integration Laboratory at JSC.

Musgrave served as a CAPCOM for the second and third Skylab missions, STS-31, STS-35, STS-36, STS-38 and STS-41. He was a Mission Specialist on STS-6 (1983), STS-51-F/Spacelab-2 (1985), STS-33 (1989), STS-44 (1991), and STS-80 (1996); and the Payload Commander on STS-61 (1993).

A veteran of six space flights, Musgrave has spent a total of 1,281 hours, 59 minutes, 22 seconds on space missions, including nearly 27 hours of EVA.

Musgrave is the only astronaut to have flown on all five Space Shuttles. Prior to John Glenn's return to space in 1998, Musgrave held the record for the oldest person in orbit, at age 61. He retired from NASA in 1997.

5.5 Guy Bluford

Guion Stewart Bluford Jr. (born November 22, 1942) is an American aerospace engineer, retired U.S. Air Force officer and fighter pilot, and former NASA astronaut, who is the first African American and the second person of African descent to go to space. Before becoming an astronaut, he was an officer in the U.S. Air Force, where he remained while assigned to NASA, rising to the rank of colonel.

He participated in four Space Shuttle flights between 1983 and 1992. In 1983, as a member of the crew of the Orbiter Challenger on the mission STS-8, he became the first African American in space as well as the second person of African ancestry in space, after Cuban cosmonaut Arnaldo Tamayo Méndez.

5.6 Valery Polyakov

Valeri Vladimirovich Polyakov born Valeri Ivanovich Korshunov on 27 April 1942) is a Russian former cosmonaut. He is the holder of the record for the longest single stay in space in human history, staying aboard the Mir space station for more than 14 months (437 days 18 hours) during one trip. His combined space experience is more than 22 months.

Selected as a cosmonaut in 1972, Polyakov made his first flight into space aboard Soyuz TM-6 in 1988. He returned to Earth 240 days later aboard TM-7. Polyakov completed his second flight into space in 1994–1995, spending 437 days in space between launching on Soyuz TM-18 and landing on TM-20, setting the record for the longest time continuously spent in space by an individual in human history.

5.7 Eileen Collins

Eileen Marie Collins (born November 19, 1956) is a retired NASA astronaut and United States Air Force colonel. A former military instructor and test pilot, Collins was the first female pilot and first female commander of a Space Shuttle. She was awarded several medals for her work. Colonel Collins has logged 38 days 8 hours and 20 minutes in outer space. Collins retired on May 1, 2006, to pursue private interests, including service as a board member of USAA.

Collins was selected to be an astronaut in 1990 and first flew the Space Shuttle as pilot in 1995 aboard STS-63, which involved a rendezvous between Discovery and the Russian space station Mir. In recognition of her achievement as the first female Shuttle Pilot, she received the Harmon Trophy. She was also the pilot for STS-84 in 1997.

Collins was also the first female commander of a U.S. Spacecraft with Shuttle mission STS-93, launched in July 1999, which deployed the Chandra X-Ray Observatory.

Collins commanded STS-114, NASA's "return to flight" mission to test safety improvements and resupply the International Space Station (ISS). The flight was launched on July 26, 2005, and returned on August 9, 2005. During STS-114, Collins became the first astronaut to fly the Space Shuttle through a complete 360-degree pitch maneuver. This was necessary so astronauts aboard the ISS could take photographs of the Shuttle's belly, to ensure there was no threat from debris-related damage to the Shuttle upon reentry.

On May 1, 2006, Collins announced that she would leave NASA to spend more time with her family and pursue other interests. Since her retirement from NASA, she has made occasional public appearances as an analyst covering Shuttle launches and landings for CNN.

5.8 Kathryn Thornton

I met Kathryn Thornton at a Houston event where she was part of a panel taking testimony on the next manned objective after the moon. I stood up and talked about the benefits of going to the asteroid belt.

Kathryn Ryan Cordell Thornton (born August 17, 1952 in Montgomery, Alabama) is an American scientist and a former NASA astronaut with over 975 hours in space, including 21 hours of extravehicular activity. She was the associate dean for graduate programs at the University of Virginia School of Engineering and Applied Science, currently a professor of mechanical and aerospace engineering.

Selected by NASA in May 1984, Thornton became an astronaut in July 1985. Her technical assignments have included flight software verification in the Shuttle Avionics Integration Laboratory (SAIL), serving as a team member of the Vehicle Integration Test Team (VITT) at Kennedy Space Center, and as a spacecraft communicator (CAPCOM). A veteran of four space flights, Thornton flew on STS-33 in 1989, STS-49 in 1992, STS-61 in 1993, and

STS-73 in 1995. She has logged over 975 hours in space, including more than 21 hours of extravehicular activity (EVA).

Thornton was a mission specialist on the crew of STS-33 which launched at night from Kennedy Space Center, Florida, on November 22, 1989, aboard the Space Shuttle Discovery. The mission carried Department of Defense payloads and other secondary payloads. After 79 orbits of the Earth, this five-day mission concluded on November 27, 1989, at Edwards Air Force Base, California.

On her second flight, Thornton served on the crew of STS-49, May 7–16, 1992, on board the maiden flight of the new Space Shuttle Endeavour. During the mission the crew conducted the initial test flight of Endeavour, performed a record four EVA's (space walks) to retrieve, repair and deploy the International Telecommunications Satellite (INTELSAT), and to demonstrate and evaluate numerous EVA tasks to be used for the assembly of Space Station Freedom. Thornton was one of two EVA crew members who evaluated Space Station assembly techniques on the fourth EVA. STS-49 logged 213 hours in space and 141 Earth orbits prior to landing at Edwards Air Force Base, California.

On her third flight, Thornton was a mission specialist EVA crew member aboard the Space Shuttle Endeavour on the STS-61 Hubble Space Telescope (HST) servicing and repair mission. STS-61 launched at night from the Kennedy Space Center, Florida, on December 2, 1993. During the 11-day flight, the HST was captured and restored to full capacity through a record five space walks by four astronauts, including Thornton. After having traveled 4,433,772 miles in 163 orbits of the Earth, the crew of Endeavour returned to a night landing at the Kennedy Space Center on December 13, 1993. Then, after Expedition 14, Sunita Williams surpassed her for woman with the most spacewalks.

From October 20 to November 5, 1995, Thornton served aboard Space Shuttle Columbia on STS-73, as the payload commander of the second United States Microgravity Laboratory mission. The mission focused on materials science, biotechnology, combustion science, the physics of fluids, and numerous scientific experiments housed in the pressurized Spacelab module. In completing her fourth space flight, Thornton orbited the Earth 256 times, traveled over 6 million miles, and logged a total of 15 days, 21 hours, 52 minutes and 21 seconds in space.

Thornton left NASA on August 1, 1996.

6.0 ISS Construction and Space Shuttle Missions

The construction of the International Space Station and the crewing of it was a major stage of manned flight into space. Here are just a few of the Astronauts who contributed to this great construction feat. There were also many Space Shuttle Missions in this era which accomplished wonderful things like repairing the Hubble Space Telescope.

6.1 David Wolf

David Wolf was a friend of mine when I lived in Houston, Texas. He took me up in his acrobatic plane where I barfed up everything. This gave me the impetus to get my own pilots license. He worked in the Medical Systems division at the time and we had a few interesting discussions about how to get into the space program

David Alexander Wolf (born August 23, 1956) is an American astronaut, medical doctor and electrical engineer. Wolf has been to space four times. Three of his spaceflights were short-duration Space Shuttle missions, the first of which was STS-58 in 1993, and his most recent spaceflight was STS-127 in 2009.

Wolf also took part in a long-duration mission aboard the Russian space station Mir which lasted 128 days, and occurred during Mir EO-24. He was brought to Mir aboard STS-86 in September 1997, and landed aboard STS-89 in January 1998. In total Wolf has logged more than 4,040 hours in space. He is also a veteran of 7 spacewalks totaling 41hrs 17min in both Russian and American spacesuits.

He also participated in the construction of the ISS.

6.2 Julie Resnick

I wrote Julie a letter about becoming an astronaut while still in college and she sent me a very encouraging reply. Later in Houston I ran into her while she was biking and I was jogging and had a short chat with her. She was a very nice person and of course I was shocked when she was killed in the Challenger explosion.

Judith Arlene Resnik April 5, 1949 – January 28, 1986 was an American electrical engineer, software engineer, biomedical engineer, pilot and NASA astronaut who died when the Space Shuttle Challenger was destroyed during the launch of mission STS-51-L. Resnik was the second American woman in space and the fourth woman in space worldwide, logging 145 hours in orbit. She was the first Jewish woman of any nationality in space. The IEEE Judith Resnik Award for space engineering is named in her honor.

Judith Resnik was accepted at Carnegie Mellon after being 1 of only 16 women in the history of the United States to have attained a perfect score on the SAT exam at the time. She went on to graduate with a degree in electrical engineering from Carnegie Mellon before attaining a Ph.D. in electrical engineering from the University of Maryland. Recognized while still a child for her "intellectual brilliance," Resnik went on to work for RCA as an engineer on Navy missile and radar projects, was a senior systems engineer for Xerox Corporation and published research on special-purpose integrated circuitry before she was recruited by NASA to the astronaut program as a mission specialist at age 28. While training on the astronaut program, she developed software and operating procedures for NASA missions. She was also a pilot and made research contributions to biomedical engineering as a research fellow of biomedical engineering at the National Institutes of Health.

6.3 Joe Allen

I met Joe while interviewing with him for a position in his space company after he left NASA. A really interesting and nice guy.

Allen was selected as a scientist-astronaut by NASA in August 1967 as a member of the second group of scientist-astronauts. He completed flight training at Vance Air Force Base, Oklahoma. He served as mission scientist while a member of the astronaut support crew for Apollo 15 and served as a staff consultant on science and technology to the President's Council on International Economic Policy.

From August 1975 to 1978, Allen served as NASA Assistant Administrator for Legislative Affairs in Washington, D.C. returning to the Johnson Space Center in 1978, as a senior scientist astronaut, Allen was assigned to the Operations Mission Development Group. He served as a support crew member for the first orbital flight test of the Space Shuttle (Columbia) in April 1981 and was the CAPCOM during the reentry phase for this mission. In addition, in 1980 and 1981, he worked as the technical assistant to the director of flight operations.

Space experience

Allen served as mission specialist on STS-5, the first fully operational flight of the Space shuttle program, which launched from Kennedy Space Center, Florida, on November 11, 1982. He was accompanied by Vance D. Brand (spacecraft commander), Col. Robert F. Overmyer (pilot), and Dr. William B. Lenoir (mission specialist). STS-5, the first mission with four crewmembers, clearly demonstrated the Space Shuttle as fully operational by the successful first deployment of two commercial communications satellites from the Orbiter's payload bay.

The mission marked the first use of the Payload Assist Module (PAM-D), and its new ejection system. Numerous flight tests were performed throughout the mission to document Shuttle performance during launch, boost, orbit, atmospheric entry and landing phases. STS-5 was the last flight to carry the Development Flight Instrumentation (DFI) package to support flight testing. A Getaway Special, three Student Involvement Projects, and medical experiments were included on the mission. A planned spacewalk by Allen and Lenoir, the first of the Space Shuttle program, was postponed by one day after Lenoir became ill, and then had to be cancelled when the two spacesuits that were to be used developed problems. The STS-5 crew successfully concluded the 5-day orbital flight of Space Shuttle Columbia with the first entry and landing through a cloud deck to a hard-surface runway and demonstrated maximum braking. STS-5 completed 81 orbits of the Earth in 122 hours before landing on a concrete runway at Edwards Air Force Base, California, on November 16, 1982.

Allen was a mission specialist on STS 51-A, which launched from Kennedy Space Center, Florida, on

November 8, 1984. He was accompanied by Captain Frederick (Rick) Hauck (spacecraft commander), Captain David M. Walker (pilot), and fellow mission specialists, Dr. Anna Lee Fisher and Commander Dale Gardner. This was the second flight of the Orbiter Discovery. During the mission the crew deployed two satellites, Canada's Anik D-2 (Telesat H) and Hughes' LEASAT-1 (Syncom IV-1), and operated the 3M Company's Diffusive Mixing of Organic Solutions experiment.

In the first space salvage attempt in history, Allen and Gardner performed spacewalks and successfully retrieved for return to Earth the Palapa B-2 and Westar VI communications satellites. STS 51-A completed 127 orbits of the Earth in 192 hours before landing at Kennedy Space Center, Florida, on November 16, 1984. With the completion of this flight Allen logged a total of 314 hours in space.

6.4 Scott Kelly

Scott Joseph Kelly (born February 21, 1964) is an American engineer, retired astronaut, and naval aviator. A veteran of four space flights, Kelly commanded the International Space Station (ISS) on Expeditions 26, 45, and 46.

Kelly's first spaceflight was as pilot of Space Shuttle Discovery during STS-103 in December 1999. This was the third servicing mission to the Hubble Space Telescope, and lasted for just under eight days. Kelly's second spaceflight was as mission commander of STS-118, a 12-day Space Shuttle mission to the ISS in August 2007. Kelly's third spaceflight was as a crewmember on Expedition 25/26 on the ISS. He arrived at the ISS aboard Soyuz TMA-01M on 9 October 2010, and served as a flight engineer until he took over command of the station on 25 November 2010 at the start of Expedition 26. Expedition 26 ended on 16 March 2011 with the departure of Soyuz TMA-01M.

In November 2012, Kelly and Russian cosmonaut Mikhail Kornienko were selected for a year-long mission to the

ISS. Their year in space began with the launch of Soyuz TMA-16M on March 27, 2015, and they remained on the station for Expeditions 43, 44, 45, and 46. The mission ended on March 1, 2016, with the departure of Soyuz TMA-18M from the station.

Scott Kelly retired from NASA in 2016. His identical twin brother, Mark Kelly, is also a retired astronaut.

7.0 Private Spacecraft

7.1 SpaceX Dragon Manned Capsule

For the first time in history, NASA astronauts have launched from American soil in a commercially built and operated American crewed spacecraft on its way to the International Space Station.

The SpaceX Crew Dragon spacecraft carrying NASA astronauts Robert Behnken and Douglas Hurley lifted off at 3:22 p.m. EDT Saturday on the company's Falcon 9 rocket from Launch Complex 39A at NASA's Kennedy Space Center in Florida.

"Today a new era in human spaceflight begins as we once again launched American astronauts on American rockets from American soil on their way to the International Space Station, our national lab orbiting Earth," said NASA Administrator Jim Bridenstine. "I thank and congratulate Bob Behnken, Doug Hurley, and the SpaceX and NASA teams for this significant achievement for the United States. The launch of this commercial space system designed for humans is a phenomenal demonstration of

American excellence and is an important step on our path to expand human exploration to the Moon and Mars." Known as NASA's SpaceX Demo-2, the mission is an end-to-end test flight to validate the SpaceX crew transportation system, including launch, in-orbit, docking and landing operations. This is SpaceX's second spaceflight test of its Crew Dragon and its first test with astronauts aboard, which will pave the way for its certification for regular crew flights to the station as part of NASA's Commercial Crew Program.

"This is a dream come true for me and everyone at SpaceX," said Elon Musk, chief engineer at SpaceX. "It is the culmination of an incredible amount of work by the SpaceX team, by NASA and by a number of other partners in the process of making this happen. You can look at this as the results of a hundred thousand people roughly when you add up all the suppliers and everyone working incredibly hard to make this day happen."

The program demonstrates NASA's commitment to investing in commercial companies through public-private partnerships and builds on the success of American companies, including SpaceX, already delivering cargo to the space station.

7.2 Boeing Starliner

NASA astronauts Nicole Mann, Michael Fincke, Suni Williams, Josh Cassada, and Eric Boe pose for a picture after a United Launch Alliance Atlas V rocket with Boeing's CST-100 Starliner spacecraft onboard was rollout out to the launch pad at Space Launch Complex 41 ahead of the Orbital Flight Test mission, Wednesday, Dec. 18, 2019 at Cape Canaveral Air Force Station in Florida. Mann, Fincke, and Boeing Astronaut Chris Ferguson are assigned to fly on Starliner's Crew Flight test, Williams and Cassada are assigned to the first operational mission of the spacecraft, and Boe is the assistant to the chief of the astronaut office for commercial crew. The Orbital Flight Test with be Starliner's maiden mission to the International Space Station for NASA's Commercial Crew Program. The mission, currently targeted for a 6:26 a.m. EST launch on Dec. 20, will serve as an end-to-end test of the system's capabilities.

7.3 Virgin Galactic Spacecraft

Mojave, California

After months of testing Richard Branson's rocket-powered plane at lower altitudes, two of Virgin Galactic's test pilots made it to the edge of space on Thursday — 51.4 miles above Earth.

For pilots Mark "Forger" Stucky and Frederick "CJ" Sturckow, looming over the flight was the memory of a tragic test flight in 2014. Virgin Galactic's first vehicle, SpaceShipTwo, ripped apart in mid-air, killing a co-pilot. Stucky spoke to CNN Business after Thursday's successful firing of the rebuilt SpaceShipTwo, called VSS Unity, to record heights at nearly three times the speed of sound. He said it was like taking a thoroughbred racehorse into a full gallop for the first time.

"Before you can race her you have to train and walk her down uneven terrain, but eventually you have to say, maybe I should race her," he said. "That's what Unity reminded me of."

The test flight was the first time Branson's space tourism startup has gone more than 50 miles above Earth. It earned both pilots commercial astronaut wings from the US government and put Virgin Galactic on track to become the first private company in the world to take paying customers to space.

Virgin Galactic had worked toward the goal since it was founded in 2004. It also marked the first crewed flight to space from US soil since the Space Shuttle retired in 2011, with Virgin Galactic beating out other well-funded competitors.

8.0 Other Notable Space Flyers

There have also been many space flyers from other countries. Here are some of the most notable ones.

8.1 Helen Sharman-1st British Astronaut

After responding to a radio advertisement asking for applicants to be the first British space explorer, Helen Sharman was selected for the mission live on ITV, on 25 November 1989, ahead of nearly 13,000 other applicants. The program was known as Project Juno and was a cooperative Soviet Union–British mission co-sponsored by a group of British companies.

Sharman was selected in a process that gave weight to scientific, educational and aerospace backgrounds as well as the ability to learn a foreign language.

Before flying, Sharman spent 18 months in intensive flight training in Star City. The Project Juno consortium failed to raise the monies expected, and the program was almost cancelled. With a view towards the flight's impact on international relations, the project proceeded under Soviet expense although as a cost-saving measure, less expensive experiments were substituted for those in the original plans.

The Soyuz TM-12 mission, which included Soviet cosmonauts Anatoly Artsebarsky and Sergei Krikalev, launched on 18 May 1990 and lasted eight days, most of that time spent at the Mir space station. Sharman's tasks included medical and agricultural tests, photographing the British Isles, and participating in a licensed amateur radio hookup with British schoolchildren. She landed aboard Soyuz TM-11 on 26 May 1991, along with Viktor Afanasyev and Musa Manarov.

Sharman was 27 years and 11 months old when she went into space, making her (as of 2017) the sixth youngest of the 556 individuals who have flown in space. Sharman has not returned to space, although she was one of three British candidates in the 1992 European Space Agency astronaut selection process and was on the shortlist of 25 applicants in 1998.

Since Juno was not an ESA mission, Tim Peake became the first ESA British astronaut more than 20 years later.

For her Project Juno accomplishments, Sharman received a star on the Sheffield Walk of Fame.

8.2 Yang Liwei-First Chinese Taikonaut

Yang Liwei (born 21 June 1965) is a major general, military pilot, and China National Space Administration astronaut.

In October 2003, he became the first person sent into space by the Chinese space program. This mission, Shenzhou 5, made China the third country to independently send humans into space.

Yang was selected as an astronaut candidate in 1998 and has trained for space flight since then. He was chosen from the final pool of 13 candidates to fly on China's first manned space mission. A former fighter pilot in the Aviation Military Unit of the PLA, he held the rank of Lieutenant Colonel at the time of his mission. He was promoted to full Colonel on 20 October 2003. According to the Youth Daily, the decision had been made in advance of his spaceflight, but Yang was not made aware of it.

He was launched into space aboard his Shenzhou 5 spacecraft atop a Long March 2F rocket from Jiuquan Satellite Launch Center at 09:00 CST (01:00 UTC) on 15 October 2003. Prior to his launch almost nothing was made public about the Chinese astronaut candidates; his selection for the Shenzhou 5 launch was only leaked to the media one day before the launch.

Yang Liwei has reported the apparition of abnormal vibrations 120 seconds after launch, he described as "very uncomfortable". As a consequence, corrective measures were swiftly taken to the design of the following CZ-2F carrier rocket for the Shenzhou-6.

Yang punctuated his journey with regular updates on his condition—variations of "I feel good", the last coming as the capsule floated to the ground after re-entry. He spoke to his wife as the Shenzhou 5 started its eighth circuit around the Earth, assuring her from space: "I feel very good, don't worry". He ate specially designed packets of shredded pork with garlic, Kung Pao chicken and eight treasure rice, along with Chinese herbal tea. In the middle of the journey, state television broadcast footage of Yang waving a small flag of the People's Republic of China and that of the United Nations inside his capsule.

State media said Yang's capsule was supplied with a gun, a knife and tent in case he landed in the wrong place.

Yang's craft landed in the grasslands of the Chinese region of Inner Mongolia at around 06:30 CST on October 16, 2003 (22:00 UTC), having completed 14 orbits and travelled more than 600,000 km. Yang left the capsule about 15 minutes after landing, and was congratulated by Premier Wen Jiabao. But the astronaut's bleeding lips seen in the official images broadcast sparked rumours of a hard landing confirmed by accounts of personnel present at the landing site.

Although the first Chinese citizen in space, Yang Liwei is not the first person of Chinese origin in space. Shanghai-born Taylor Wang flew on Space Shuttle mission STS-51-B in 1985. Wang, however, had become a United States citizen in 1975.

Yang visited Hong Kong on 31 October 2003, holding talks and sharing his experiences during a six-day stay in the territory. The visit coincided with an exhibition that featured his reentry capsule, spacesuit and leftover food from his 21-hour mission. On November 5, he travelled to Macau. On 7 November, Yang received the title of "Space Hero" from Jiang Zemin, the Chairman of the PRC Central Military Commission (CMC). He also received a badge of honor during a ceremony at the Great Hall of the People. Russia awarded him the Gagarin medal. The Chinese University of Hong Kong has given Yang an honorary doctorate.

Important Astronauts, Cosmonauts, and Other Spaceflight Personalities

8.3 Koichi Wakata-First Japanese Astronaut

Koichi Wakata born 1 August 1963 is a Japanese engineer and a JAXA astronaut. Wakata is a veteran of four NASA Space Shuttle missions, a Russian Soyuz mission, and a long-duration stay on the International Space Station.

During a nearly two-decade career in spaceflight, he has logged more than eleven months in space. During Expedition 39, he became the first Japanese commander of the International Space Station. Wakata flew on the Soyuz TMA-11M/Expedition 38/Expedition 39 long duration spaceflight from 7 November 2013 to 13 May 2014. During this spaceflight he was accompanied by Kirobo, the first humanoid robot astronaut.

Important Astronauts, Cosmonauts, and Other Spaceflight Personalities

8.4 Rakesh Sharma-First India Cosmonaut

Wing Commander Rakesh Sharma, AC (born 13 January 1949) is a former Indian Air Force pilot who flew aboard Soyuz T-11 on 3 April 1984 with the Interkosmos program. He is the only Indian citizen to travel in space, although there have been other astronauts with an Indian background who were not Indian citizens.

An alumnus of the 35th National Defence Academy, Sharma joined the Indian Air Force as a test pilot in 1970 and progressed through numerous levels where in 1984 he was promoted to the rank of squadron leader. He was selected on 20 September 1982 to become a cosmonaut and go into space as part of a joint program between the Indian Air Force and the Soviet Interkosmos space program.

In 1984, Sharma became the first Indian citizen to enter space when he flew aboard the Soviet rocket Soyuz T-11 launched from Baikonur Cosmodrome in the Kazakh Soviet Socialist Republic on 3 April 1984. The Soyuz T-11 spacecraft carrying cosmonauts including Sharma docked and transferred the three member Soviet-Indian international crew, consisting of the ship's commander, Yury Malyshev, and flight engineer, Gennadi Strekalov, to the Salyut 7 Orbital Station. Sharma spent 7 days, 21 hours, and 40 minutes aboard the Salyut 7 during which his team conducted scientific and technical studies which included forty-three experimental sessions. His work was mainly in the fields of bio-medicine and remote sensing. The crew held a joint television news conference with officials in Moscow and then Indian Prime Minister Indira Gandhi. When Gandhi asked Sharma how India looked from outer space, he replied, "Sare Jahan Se Accha" (the best in the world). This is the title of a patriotic poem by Iqbal that had been written when India was under British colonial rule, that continues to be popular today. With Sharma's voyage aboard Soyuz T-11, India became the 14th nation to send a man to outer space.

Sharma retired as a wing commander and later joined Hindustan Aeronautics Limited (HAL) in 1987, serving as the chief test pilot in the HAL Nashik Division until 1992, before moving on to Bangalore to work as HAL's chief test pilot. Sharma retired from flying in 2001.

9.0 Summary

It was really interesting researching all of these Astronauts, Cosmonauts, and more. I learned a lot of things I didn't know including the backgrounds of these space flyers, their previous jobs, and what they did at their space agencies in training or before they became astronauts.

Space flyers are often seen as national heroes, and many of them deserve that accolade for their bravery in trying new spacecraft and rockets.

However, there are many lesser known space flyers who also accomplished many unique things.

Lastly, we should remember the hundreds of thousands of designers, workers, support personnel and more who make it possible for the select few to fly into space.

Hopefully, space travel is becoming more affordable so that everyone will have a chance to go into space in the future.

Martin K. Ettington

September 2020

Important Astronauts, Cosmonauts, and Other Spaceflight Personalities

10.0 Bibliography

1. US Astronaut Hall of Fame Inductees. *https://en.wikipedia.org/wiki/United_States_Astronaut_Hall_of_Fame#Inductees.* [Online]

2. Famous Astronauts. *Search of Google for "Famous Astronauts".* [Online]

Important Astronauts, Cosmonauts, and Other Spaceflight Personalities

11.0 Index